Design and Karting

Schools Council Design and Craft Education Project

Edward Arnold

Contents

Preface   vii

Introduction   1

1 Introduction and Rationale   4

2 Organization and Operation   9

3 Designs and Specifications   19

4 A Case Study   32

5 Kart-ways   42

6 Karting Questionnaires   53

Bibliography and Further Reading   68

Acknowledgments

We would like to thank The National Schools' Karting Association and The Hertfordshire Schools' Karting Association for their advice and assistance in the preparation of this book.

Preface

This volume and its associated publications represents part of the outcome of the Schools Council development project in Design and Craft Education. The project commenced in 1966 when the Schools Council generously financed a feasibility study to explore the possibilities of development that might spring from the existing school curricula in woodwork, metalwork and related activities with particular reference to the needs of the older age groups in the secondary schools. The Schools Council decision was in response to the enthusiastic initiatives of a large group of teachers in schools, colleges and universities who, with heads, advisers and administrators had met together to formulate proposals on a number of occasions at Leicester throughout 1965.

The feasibility study outlined a range of promising developments that were already taking seed in the schools and, in response to the report of this study (published as Schools Council Working Paper No. 26 *Education Through the Use of Materials*) the Council made a series of further grants to enable a major development study to take place at the University of Keele. Overall the project continued for six years with a total grant amounting to some £73,000.

Working with me at Keele for all or some of the time have been Louis Brough, Russell Hall, Allan Pemberton, Denis Taberner and our secretary Mrs Gillian Hope. But the efforts of our team at Keele have been hugely augmented by the work of the teachers in the schools and of their students. Most crucial has been the work of those in the trial schools in England and Wales and, later, in Northern Ireland and the British schools in Germany where the exhaustive initial work was undertaken, much of it initiated by the teachers themselves. In this we were particularly helped by the work of a number of teacher coordinators who were relieved of some of their normal teaching duties for varying periods by their local education authorities in order to participate in the Project. These were Mr D. C. Crook, Mr R. W. Draycon, Mr G. Hughes, Mr D Hunt, Mr T. Maguire, Mr G. Martin, Mr B. S. Nicholson, Mr J. D. Parker, Mr Paulden, Mr D. V. Pyatt, Mr J. Richards, Mr J. Ridge, Mr P. L. Small, Mr E. Swan, Mr A. Wagstaff and Mr R. Waterhouse.

But in addition to the work in the trial schools many other teachers in the design subjects in other schools in various parts of Britain and the rest of the world gradually came to be involved in the general exchange of ideas that sustained and enhanced the development. A notable part of the project was also played by the specialist advisers and inspectors of the local education authorities in whose schools we were working. But over and above the contribution of the design specialists there was also the more general contribution of other teachers in the schools, the head teachers and local authority officers and the members of the various examination boards, most notably those of the North Western Secondary Schools Examinations Board who joined us in introducing entirely new examinations in design education. These contributions were at all times abundant and generous and their importance is reflected throughout the publications as is the outstanding assistance that was received from members of the Inspectorate, from the officers of the Schools Council, from a wide range of industrial and commercial organizations and from our publishers. Much of this support reached the team through the representative advisory committee which, under the chairmanship of Professor G. H. Bantock, effectively guided the project through its existence.

The varied publications of the project are only a part of its outcome. There are many others including the development of new strategies of examination in the design subjects, the in-service training of design teachers and the by now self-perpetuating development of ideas in the schools. These are at least of equal importance. Indeed, we hope that the publications are seen not as an end but as a means to such continuing development throughout the whole field of design education.

JOHN EGGLESTON

Introduction

For several years a group of lecturers, teachers and advisors meeting at the University of Leicester attempted, through conferences and working parties, to explore new approaches to the teaching of handicrafts at the secondary level. This work showed, among other things, an urgent need for development studies of this area of the curriculum, particularly in the context of the raising of the school-leaving age. Support for such a project was widely expressed and it was felt that the possibility of establishing a feasibility study, from which a substantive research and development project might be developed, should be explored. A steering committee was elected to formulate proposals and present them to the Schools Council. A list of members of the original committee and those who joined subsequently is to be found in Appendix 1.

The proposals were accepted and funds were made available for a one-year feasibility study which started on 1 April 1967. A report of the study, published as Schools Council Working Paper 26, *Education Through the Use of Materials*, describes the work undertaken, identifies a number of curriculum growth points and outlines proposals for further investigation. Arising out of this work a three-year substantive research and development project, funded by Schools Council, was established at the University of Keele (where the Project Director had been appointed to the Chair of Education) and work started on 1 April 1968. Subsequently the Project was extended for a further two years to develop in-service teacher training and to explore ways in which the work might be related to external examinations.

Education Through Design and Craft describes the work of the substantive project. It looks at the factors that have led to new developments in the materials subjects and examines the role these developments have to play in education in a changing society. Particular attention is paid to aims and objectives and examination and assessment.

At the heart of the Project's work has been the development of problem-solving approaches suitable for use in secondary schools. These approaches

1

and their development and implementation at various levels throughout the school are the subject of *Materials and Design: A Fresh Approach*, a book written for teachers in service and teachers in training. It shows how students can be encouraged to identify design problems, investigate them, produce and realize solutions and finally evaluate the end-products.

Design for Today builds upon the design approaches examined in *Materials and Design: A Fresh Approach* and looks more specifically at some of the ways in which they may be applied in activities relevant to the needs and interests of students of a wide range of abilities.

Five filmstrips have been prepared to introduce specific topics that may be used as a basis for design work. Commentaries and frame reproductions appear in *Design for Today*. The filmstrip titles are: *Houses and Homes*, *Value for Money* (design and consumer discrimination), *Helping Out* (design and community action), *Design and the Environment*, and *Playthings* (the design of play equipment).

You Are a Designer speaks directly to the student, introducing him to the problem-solving approach, first in a general, descriptive way and then by listing and detailing the factors that may need to be considered at various stages during a design process and suggesting a range of key questions that he may need to ask. Thus the book has two functions, being useful both as an introduction to designing and as a 'memory-jogger' whenever a design process is being put into action.

Also aimed at the student is *Connections and Constructions*, a resource book designed to assist the development of appropriate fabrication techniques when designing and making products in a wide range of hard and soft materials.

A series of filmstrips has been produced to introduce certain principles essential to much design and craft work. Fourteen subjects—including shape, form, colour, texture, ergonomics, materials and tools—are covered in nine filmstrips. Commentaries and frame reproductions of this series appear in *Looking at Design*.

It was felt that certain specialized areas of work required more detailed attention than was possible in the above publications. For this reason two supplementary teachers' books, in addition to this present volume, have been prepared, *The Creative Use of Concrete* and *Designing with Plastics*. In each case the emphasis is, of course, on a design-based approach. Introductory filmstrips *Kart-ways* and *Designing with Concrete* support two of these publications.

In addition, the National Council for Educational Technology has, in association with the Design and Craft Education Project, produced a film, *Design With a Purpose*, which looks at the work of a group of fifteen-year-old students involved in designing and making play equipment for infant school children as part of an experimental Mode 3 CSE course. The film may be used by teachers' groups as a focus for discussion and development, and for students it offers an introduction to the requirements, and flexibility, of a design process. *Design With a Purpose* is distributed by the EFVA and is available from the National Audio Visual Aids Library, Paxton Place, Gipsy Road, London, SE27 9SR.

1
Introduction and Rationale

The role of the motor vehicle in modern society is an important and all-pervading one. Youngsters now at school have been familiar with motor vehicles all their lives. In many cases their interest lies not in the vehicle's prime function, that of transport, but more often in the sensations to be gained from speed and the skilful control of a moving machine.

For some years teachers have built upon this interest by including motor vehicle work in the craft curriculum and, more recently, by developing specific courses that include instruction in driving.* Associated with these developments has been the growth of interest in karting as a school activity. But, although much has been achieved, so much potential remains that the activity may still be said to be in its infancy. The intention of this book is to discuss the work that is being done and to outline some of the possibilities for future development.

The appeal of karting

The recreational aspects of karting can be of great value. Karting introduces students to an exciting sport and may well develop interests that they can carry over into adult life. It is the cheapest form of motor sport and allows drivers to travel fast and develop their skills under relatively safe conditions.

As with many other sports, the equipment required can be purchased. But a commercially built kart and trailer would, of course, be far beyond the financial resources of the average fourth- or fifth-year student. Fortunately, many schools have discovered that not only is it possible for students to build karts effectively and cheaply, but also that the process of designing and building a kart generates enormous interest.

*See *Design for Today*, Chapter 3.

4

Designing and driving

In the adult world ownership of a motor vehicle is an increasing necessity. Many schools are developing courses in motor maintenance which often include some form of instruction in driving. But this aspect is usually confined to driving a vehicle of very limited performance in the restricted space of the playground. Karting can act as a valuable extension to this type of course, offering the student experience in both maintenance and driving. Such things as skid correction and reaction testing are basically similar in cars and karts, but are much safer practised in the latter.

Designing and building a kart provides a wide range of opportunities for students to explore current trends in the functional design of motor vehicles. It involves many engineering processes that demand high standards of precision. Students are therefore presented with real problem-solving situations and realize that the efficiency and safety of the finished kart will depend on their solutions and the way they are put into practice.

1.1 Engineering processes may require high standards of precision.

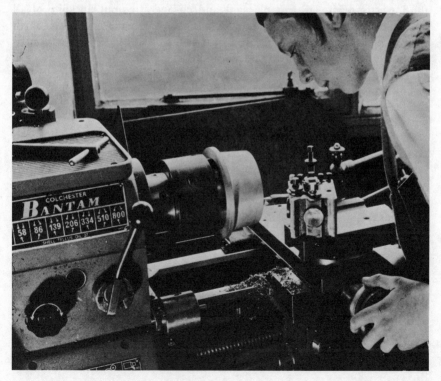

Social education

Finally, karting offers an opportunity for the development of social awareness. Students need to work in teams to achieve a common end. They experience the need for co-operation, the division of labour, the acceptance of personal responsibility for work standards and for disciplined conduct when driving on a track.

In addition, there is enormous potential in the social contacts made between teachers, students and parents for the extension of the students' social education. This aspect may be expanded in a situation where it is possible for a group of schools to join together and set up some kind of organization to finance and arrange a combined programme of track events.

The individual student

Like all school activities, karting has something rather special to offer the individual boy or girl who identifies with it. He or she gives more to the activity and so is able to derive from it something of the kind of experience necessary to the development of a mature human being.

One example of this is the case of a boy who spent his school career in the lowest stream. He was pleasant and co-operative but, in spite of all efforts, his teachers expected him to leave school barely literate. In his fourth year he became very much involved with karting and made a major contribution to the development of a school kart.

He also saved enough money to buy himself a second-hand kart so that he could continue the sport when he left school. His decision to stay on into the fifth year was largely the result of his desire to pursue his interest in karting.

His parents said that this was the first thing he had ever really been good at—and he really was good! He had gained immeasurably from the self-confidence associated with his success. He qualified for his county kart association's driving licence—an achievement that was the source of considerable satisfaction to him—and always showed responsibility in his track activities.

Aims and objectives of karting in schools

Some of the educational implications of karting in schools have been dealt with in the preceding pages. Some more specific aims and objectives of this work are outlined in diagrammatic form in figure 1.2. These fall into three general areas, which are not, however, exclusive.

6

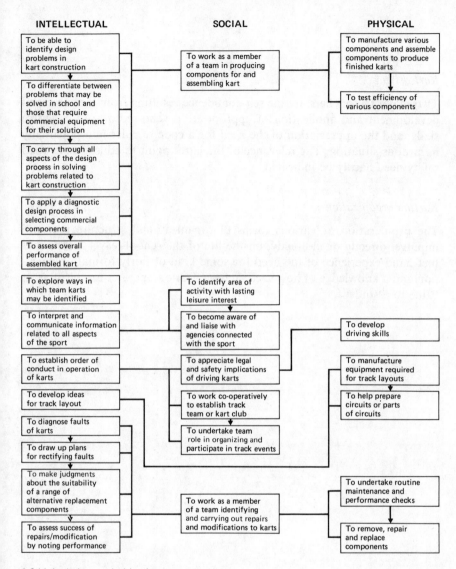

INTELLECTUAL

To be able to identify design problems in kart construction

To differentiate between problems that may be solved in school and those that require commercial equipment for their solution

To carry through all aspects of the design process in solving problems related to kart construction

To apply a diagnostic design process in selecting commercial components

To assess overall performance of assembled kart

To explore ways in which team karts may be identified

To interpret and communicate information related to all aspects of the sport

To establish order of conduct in operation of karts

To develop ideas for track layout

To diagnose faults of karts

To draw up plans for rectifying faults

To make judgments about the suitability of a range of alternative replacement components

To assess success of repairs/modification by noting performance

SOCIAL

To work as a member of a team in producing components for and assembling kart

To identify area of activity with lasting leisure interest

To become aware of and liaise with agencies connected with the sport

To appreciate legal and safety implications of driving karts

To work co-operatively to establish track team or kart club

To undertake team role in organizing and participate in track events

To work as a member of a team identifying and carrying out repairs and modifications to karts

PHYSICAL

To manufacture various components and assemble components to produce finished karts

To test efficiency of various components

To develop driving skills

To manufacture equipment required for track layouts

To help prepare circuits or parts of circuits

To undertake routine maintenance and performance checks

To remove, repair and replace components

1.2 Linked aims and objectives.

Kart design and construction

The educational objectives of the design process have been discussed in some detail in *Materials and Design: A Fresh Approach* and these will apply to all design work in schools. Kart design does, however, lend itself to the emphasis of certain of these objectives. A kart is a fairly complex machine whose performance will depend on the care and skill that has been applied by each member of a team of students. The effects of poor design or craftsmanship, or lack of co-ordination, are likely to be very apparent in the finished vehicle.

Kart driving

Two major factors here are the self-confidence resulting from the successful development and application of appropriate psycho-motor and cognitive skills, and the appreciation of the need for a code of rules in a potentially hazardous situation. The relevance of the latter point to education in road safety need hardly be indicated.

Karting organization

The proliferation of various forms of organizational structure that will impinge, directly or indirectly, on the life of the school-leaver means that first-hand experience of the need for some kind of fairly formal organization and a knowledge of how such an organization operates can be of great value to a student.

2
Organization and Operation

In many kinds of teaching it could be argued that the motivation of the students makes the greatest demands upon the teacher. Karting, however, has instant appeal. Indeed, most teachers have found that the first problem was to control the enthusiasm engendered by the announcement that karting was to begin in their school. From that point on the process is usually one of harnessing the enthusiasm to fulfil educational requirements.

The central issue here is that the students' enthusiasm is initially concerned with the actual driving of the kart rather than the long and complicated business of designing and building one. The teacher's role is to persuade his students to combine this thrill with kart development.

The undoubted values of kart driving have already been discussed; nevertheless, there is an obvious danger if students are allowed to see the enterprise as a kind of rodeo or fairground activity. At best, school karting has resulted in many examples of first class engineering design, development and construction. At worst, the activity could degenerate into a noisy, expensive and dangerous pursuit with more frustrations than thrills.

Maintaining enthusiasm

Motivation at the outset, then, is not difficult. It may, however, be thought worth while to inject some kart driving experience at an early stage to maintain enthusiasm. Some teachers recommend a purchase of a second-hand kart for initial use. Alternatively, it may be possible to share the resources of a school where activities are already well developed. In one county a kart made at the local Polytechnic has been made available to a number of schools to help stimulate initial interest. Such stimulus can also help to revitalize flagging enthusiasm when work reaches a difficult phase.

In some parts of the country there are tracks where experienced drivers

9

can be seen handling very powerful machines with great skill. A visit to such a track can also prove a useful enthusiasm booster.

Complementary to this booklet is a filmstrip entitled *Kart-ways* which is intended to act as a basis for discussion at the beginning of a kart project. Commentary and frame reproductions appear in Section 5. The detailed frame-by-frame notes draw attention to many points that should be raised at the outset. Individual frames can be used later on for a more specific analysis of particular areas of design.

Reproduced in Section 6 is a series of eight questionnaire sheets for students, which have been designed to act as an introduction to eight basic areas of investigation that must be tackled by anyone setting out to design and construct a kart. These are:

1 the basic frame;
2 wheels and braking;
3 engines—types and methods of mounting—exhaust systems;
4 axles—bearings and drive mechanisms;
5 steering;
6 the fuel system;
7 seating and controls;
8 social organization.

Design group organization

How do we begin the design process? Experienced teachers will be aware that the extent to which students can contribute to the design of their work depends not only on their intelligence and experience, but also on the working conditions. For example, the size of the group and their relationship with the teacher can have a considerable effect on the quality of their work.

The experience of many teachers has shown, however, that even those students of low academic ability can, when faced with a tangibly expressed practical problem, make a greater contribution than might at one time have been supposed in the fields of design, problem-solving and decision-making.

These areas of work are discussed at some length in *Materials and Design: A Fresh Approach*. Many of the points dealt with are applicable to all forms of design work. A number of additional factors that may have to be considered when organizing a kart design group are detailed on the following pages.

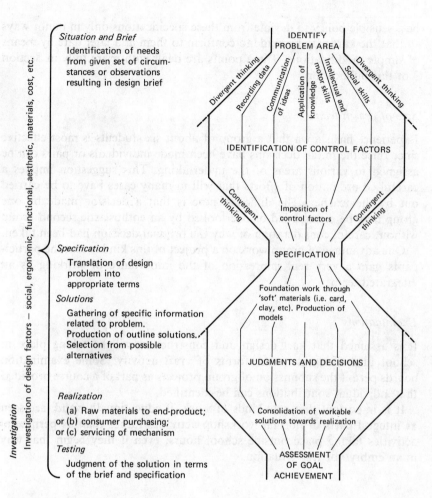

Investigation of design factors — social, ergonomic, functional, aesthetic, materials, cost, etc.

Investigation

Situation and Brief

Identification of needs from given set of circumstances or observations resulting in design brief

IDENTIFY PROBLEM AREA

Divergent thinking

Recording data

Communication of ideas

Application of knowledge

Intellectual and motor skills

Social skills

Divergent thinking

IDENTIFICATION OF CONTROL FACTORS

Convergent thinking

Imposition of control factors

Convergent thinking

Specification

Translation of design problem into appropriate terms

SPECIFICATION

Solutions

Gathering of specific information related to problem. Production of outline solutions. Selection from possible alternatives

Foundation work through 'soft' materials (i.e. card, clay, etc). Production of models

JUDGMENTS AND DECISIONS

Realization

(a) Raw materials to end-product; or (b) consumer purchasing; or (c) servicing of mechanism

Consolidation of workable solutions towards realization

Testing

Judgment of the solution in terms of the brief and specification

ASSESSMENT OF GOAL ACHIEVEMENT

2.1 A design process.

Constraints

It will be for the individual teacher to determine what constraints are to be placed on the work at the design brief stage to enable the students to make decisions at an appropriate level. He must also make clear any other limitations arising from economic factors, availability of equipment and, of course, National Schools Karting Association (NatSKA) specifications if these are adopted.

Some schools have built first karts which did not conform to RAC or NatSKA specifications, particularly with regard to wheels or engines (scooter wheels, for example, are sometimes readily available). But it may

11

be a sensible policy to deviate from these specifications only in minor ways so that the kart may be made to conform to them at a later date by means of simple modifications. These points are dealt with more fully in Section 3 of this publication.

Size of design teams

Experience has shown that a group of about six students is most effective since once the initial decisions have been made individuals or pairs can be assigned to various areas of the undertaking. This suggestion implies a careful co-ordination of efforts that will in many cases have to be carried out by the teacher. The difficulty here is that a decision made by one group may be disregarded or overlooked by an enthusiastic second group without careful consideration of why the original decision had been taken.

One advantage of group work on a project of this kind is that the participants gain a very real impression of the importance of working as an integrated team.

Examinations

It is assumed that kart design and construction will be taking place in school time alongside other forms of craft activity. Some examination boards permit the submission of group projects as part of a course provided that individual contributions can be identified.

It is important that new kinds of work, if educationally valid, are seen as integral parts of normal workshop activities rather than as peripheral activities taking place outside school hours, even if they begin that way in an embryo or trial situation.

2.2 A karting event.

Karting clubs and events

The appearance of a kart in a school is likely to produce a flood of enquiries about how soon everyone (and the enquirer in particular) will be able to drive and take part in track events. This enthusiasm can perhaps best be channelled by forming a kart club, on the understanding that members will graduate towards driving after performing the necessary kart construction jobs and doing maintenance work.

It is important to stress that club membership is by no means just a matter of driving an exciting vehicle at someone else's expense. Indeed, many school clubs allow those students who have contributed most work to have first place in the queue for driving the kart.

A school kart club

Karting is expensive, even if a school already owns a kart. Fuel may well cost £1 per kart per week in the season, and tyres need replacing at a surprising rate at a cost of about £5·00 a time. Chains, too, wear rapidly and funds will always be needed to build the next and better kart.

A club therefore requires a regular income if it is to survive. Some headmasters are willing to make grants to support the activity, although such assistance is more readily available to start an activity than to keep it going. Parents' associations or the parents of children involved in karting activities may well be a source of support.

Circumstances vary from school to school, but where it is possible for students to take office in the club this should be encouraged. The value of such experience has been mentioned in the section of this book dealing with aims and objectives.

County school's karting associations

At the time of writing at least four counties were operating karting associations with support from the local education authority. One function of these associations is to arrange joint karting events. These are more than just a challenge and a goal for students; they also provide a useful meeting ground for parents and teachers.

In addition, county associations have already given a good deal of support to newcomers to the activity. The more highly developed karting schools can help others with equipment and advice, and the bringing together of teachers and students with shared interests does nothing but good.

It may be possible to arrange financial assistance at county level through a schools' karting association. Some local authorities have seen the educational potential of karting clearly enough to make cash available to be-

2.3 A transport trailer for karts.

ginners, although the continuing expense of maintenance will obviously have to be borne by local funds.

Associations and the legal side

In the event of an accident it seems likely that a better case could be made out in law if activities were under the jurisdiction of a well-run county organization. One county's karting association, for example, provides for scrutinizing vehicles, and driver training and examination as well as having established a code of practice for track events based initially on the RAC Kart Regulations and later on those of NatSKA.

Most students accept the necessity for regulations at track events, feeling that this is the way things are done in motor racing anyway. Penalties for infringements are laid down by the county associations but the most severe found necessary up to the time of writing has been a firm reprimand by a member of staff or the removal of the offender from the track for a short period.

The karting event

A karting event can take place at a number of levels; some schools have the opportunity to use tracks of international standard, others use county tracks, while much karting takes place on the school playground. Nevertheless, at whatever level an event is organized it is extremely important that due regard is paid to safety. In this connection reference should be made to the handbook published by the National Schools Karting Association (see bibliography).

The reader will note that throughout this booklet the word 'event' has been used rather than 'race'. Racing as such was not permitted since insurers generally regarded it as too dangerous. However, after lengthy negotiations with the RAC, NatSKA have been able to organize race meetings for themselves and member Counties as long as the regulations conform to the NatSKA Handbook. Further information on this point is available from Hon. Secretary, NatSKA: see Bibliography and Further Reading.

While parents will frequently visit karting events and play a valuable part in the organization, it is well to determine at the outset who is to be ultimately responsible for the various activities. Authority at the event must be in the hands of the track officials and, when necessary, it should not only be exercised, it must clearly be seen to be exercised.

The track layout

The track layout (figure 2.4) should include a lane for karts to enter the track under the track marshall's direction. A lay-by ought to be provided at the end of this lane in which a kart which fails to start can shelter until it is removed by the track stewards. Similarly, drivers wishing to leave the track must be provided with a safe exit lane. They should be required to raise their hand to signal such an intention.

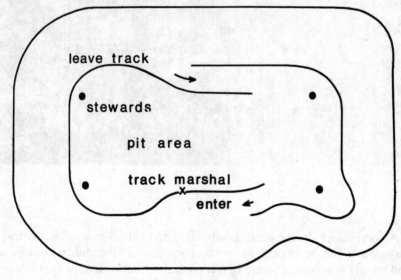

2.4 Track layout.

The ideal track has a concrete or tarmac surface; a school yard or netball court may often be suitable. Hard-packed dirt, sand or shale is hard on chains and bearings. Playing fields should also be avoided if possible; the soft surface may break some types of chassis and difficulties may be encountered in starting Class 100 engines and in cornering when the grass is wet. In addition, treaded tyres must be used and the grass surface may be badly damaged.

The pit area should contain first aid equipment and a fire extinguisher as well as kart maintenance gear. If play pitches are used good relations with the PE staff can be preserved by persuading karters to provide some form of ground covering at re-fuelling points. And kart chains should not be too liberally oiled.

2.5 The pit area.

A compulsory dinner break is advisable and rules about when the track may or may not be crossed by spectators and pit staff should be drawn up and strictly observed. The use of 'pit passes' for authorized individuals is a good idea.

Safety factors

Karting, like activities such as yachting or climbing, is potentially dangerous. But provided that those in charge are aware of the dangers and take steps to reduce them mishaps need not occur. The greatest single risk is that a kart may collide with a stationary object such as a stalled kart. Obviously it is best to lay out the track on clear ground, but if play pitches have to be used old car tyres—often obtainable from local tradesmen—can be placed around any obstructions.

Whatever additional safety precautions are taken much must depend upon track organization and discipline. Drivers must have a responsible attitude and be familiar with and obedient to the signals of the track marshal. Overturning of karts is unusual, and indeed is almost impossible.

Difficult corners should be guarded by tyres three high and secured to each other by rope. Crash helmets and goggles must be worn, together with protective overalls and gloves. It is important that teachers set a good example in these matters.

2.6 Karts at start—note the protective clothing.

Insurance

Where karting is organized as a school activity the responsibility for the safety of the participants rests with the teachers. It is well for the individual to make certain that his professional organization affords him the necessary insurance cover. Such cover normally exists provided that all reasonable precautions are in operation. Where schools join forces for events it is vital that teachers understand their responsibilities towards their own and other students. In this respect the situation is very much the same as exists in such sports as swimming, climbing, sailing and canoeing.

Some LEAs provide cover for their employees against liability for injury but it must be appreciated that this only operates if all reasonable precautions are being taken. The teacher who operates outside what might be considered a reasonable level of discipline, reliability of vehicles, and so on, does so at his own risk. It is important for the teacher involved in karting to ascertain the legal position with his own local authority and, naturally, to get the support of his headmaster. He should also ensure that parents are aware of karting activities and have given permission for their sons or daughters to participate, although he must realize that their permission does not absolve him from his professional responsibilities.

3
Designs and Specifications

The notes in this section are intended to give initial guidance to a teacher considering the introduction of karting in his school. Books dealing in depth with the technical side of karting are listed in the bibliography.

The control of motoring events in the UK is organized by the RAC and their regulations covering the design and construction of karts might prove useful, particularly where students wish to graduate towards out-of-school club activities. For development within the school, however, it is recommended that the regulations published by the National Schools Karting Association (NatSKA) are adhered to.

The NatSKA regulations

Frames, wheels and brakes

Chassis. Generally of safe and adequately strong construction and not including any components of a temporary character.

Wheel base. Maximum 50″ (127 cm), minimum 40″ (101 cm). Maximum overall length of kart 78″ (198 cm).

Track. Minimum two-thirds wheel base, maximum 45″ (114 cm).

Height. Maximum 24″ (60 cm) from ground level, excluding anti-roll or safety bar.

Frame. Sound construction without any type of structure over the driver's feet or legs. Bumper bars integral with the chassis must be provided at the front and rear. They must not extend beyond inner rims of wheels, neither must they be less than 12″ (30 cm) in length with face parallel to axles. At front and rear, no projections beyond the bumpers except wheels and tyres.

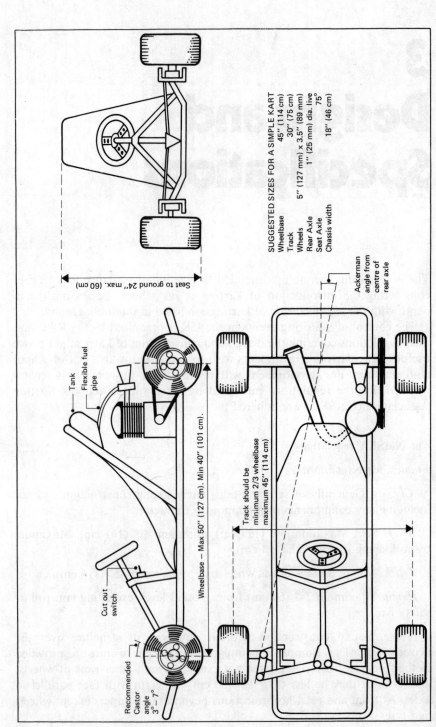

SUGGESTED SIZES FOR A SIMPLE KART

Wheelbase	45'' (114 cm)
Track	30'' (75 cm)
Wheels	5'' (127 mm) x 3.5'' (89 mm)
Rear Axle	1'' (25 mm) dia. live
Seat Axle	75°
Chassis width	18'' (46 cm)

Seat to ground 24'', max. (60 cm)

Tank

Flexible fuel pipe

Cut out switch

Recommended Castor angle 3° – 7°

Wheelbase – Max 50'' (127 cm). Min 40'' (101 cm).

Track should be minimum 2/3 wheelbase maximum 45'' (114 cm)

Ackerman angle from centre of rear axle

3.1 NatSKA regulations are indicated on this drawing.

Tyres. Pneumatic tyres with conventional tubes. Maximum diameter 17″ (43 cm), minimum 9″ (23 cm).

Wheels. Ball or roller type bearings. Bearings also for live axles.

Brakes. Foot-operated on both rear wheels, all classes. Drum or disc.

Engines, transmission and fuel supply

Engines. Series production readily available through commercial channels. No supercharging.

Capacity and transmission. Maximum engine capacity for Class 100 not to exceed 100 cc. Maximum capacity for Class 210 not to exceed 210 cc. Class 100 to be direct drive. Class 210 to have variable ratio transmission and capable of unassisted standing start.

Exhaust. Gases to rear. Silencers must be fitted. Must not project beyond chassis.

Steering. Direct or geared. Adequate construction.

Fuels and oils. Commercial types. No octane additive allowed except UCL.

Fuel and oil containers. Any material, provided they are leakproof and securely mounted. Must not project beyond chassis.

Fuel pipes and feeds. Flexible or with flexible inserts. Gravity feeds must have tap easily operated by gloved hand to turn off fuel. Fuel vents must not hazard others.

Chain guards. Designed to prevent injury should chain break. No uncovered exposed sprockets, even if not in use.

Seating, controls and weight

Seating. To give adequate back and sideways support to the driver.

Bulkhead. Metal or fire-resistant material of adequate strength as seat back. Seated driver must not come into contact with the exhaust system.

Flooring. Driver's feet must not come into contact with the track.

Control pedals. Should not protrude beyond the foremost chassis component, even when depressed.

Throttle. Foot-operated. Throttle must close completely when pedal is released or if cable breaks.

Ignition. Cut-out switch in front of driver easily operated by gloved hand from the driving position.

Means of starting. If push start, firm handhold for pushers.

Number plates. Visible front and both sides. White lettering on background colour according to class. No metal, must be flexible, 8″ (20 cm) minimum diameter, vertical to 45 degrees.

Design pointers

This section has been provided to give a teacher with little or no experience of karts some idea of the potential of kart design and the ways in which it could be tackled. It has been compiled as a result of the experience of several schools which have successfully introduced karting into their curricula.

Investigation

The initial investigation required by the design process may be broken down into eight areas. In the early stages each area may be investigated by an individual student or a small group or design team. As the investigation progresses an increasing amount of consultation and co-operation between the groups will be necessary. These areas of investigation correspond to the set of eight students' questionnaire sheets on kart design which are reproduced in Section 6.

Frame design

Schools have experimented with frame construction in a variety of materials but the consensus of opinion seems to favour round steel tube ERW/SB (electric resistance welded/semi-bright), especially where welding equipment is available. Almost essential when using this material is a tube bender with tools to handle the tube measured in OD sizes. At the time of writing this tube is only available in imperial dimensions but no doubt a gradual change towards metrication will take place. We understand that metric formers and guides are available for the Hilmor K3A tube bending machine used in many schools.

A simple kart frame can be made using a $1'' \times \frac{1}{8}''$ (25 mm × 3 mm) square mild steel tube in straight lengths bolted together with HT bolts. The more difficult types use $1''$ (25 mm) diameter × 14 G ERW tube with bronze welded joints.

There is, of course, no reason why more recently introduced materials cannot be experimented with, indeed, this can be a most interesting and valuable exercise. Standard regulations covering kart design do not preclude the use of a frame made entirely of glass-fibre. This material, perhaps combined with carbon fibre, could result in a new impetus being given to frame design.

The initial chassis drawing is best tackled on a large piece of hardboard, old carton material, or bituminous-lined building paper spread out on the floor. In this way the drawing can be chalked out, modified if necessary and then stored away for future reference. Care is, of course, necessary when offering-up hot tubular structures for checking!

The tendency in recent years has been for kart frames to approach the smaller end of the range permitted by the NatSKA regulations. Some schools are inclined, when starting, to make karts rather larger than necessary; perhaps to ensure stability.

Wheels

For the karting enthusiast outside school the acquisition of wheels is a matter of selecting from the many available good designs, perhaps the most advanced being made of nylon and fitted with tapered roller bearings. But for a school with casting equipment and a reasonably large lathe (e.g., a Colchester 'Bantam') there is a valuable field of design and development to be explored in wheel production.

Each of the front wheels usually runs on two bearings. Tapered roller bearings are best but ball bearings are cheaper. Whichever type is used care should be taken to use sealed bearings or to fit suitable cover plates to exclude grit.

The two halves of the rear wheel are different. The inner half is bored to a slide fit on the axle (nominally 1″ (25 mm)) while the outer half is bored to suit a $\frac{3}{4}$″ (19 mm) diameter portion of the axle. The difference between the two diameters is to prevent axial movement when the axle nut has been tightened.

There are many aspects of design to be considered. Figures 3.2 and 3.3 are provided as guides, not as the last word in design but to emphasize some of the design details that ought to be taken into consideration. These drawings are the result of work in one trial school which took as its brief the production of a design requiring the minimum amount of machining.

It is suggested that initially a common pattern is made to cast both front and rear wheels, provided that both use 5″ (127 mm) tyres. Later development work as students gain experience will no doubt produce different patterns for front and rear wheels.

With some types of wheels access to the valve can prove troublesome. Design teams should pay particular attention to this problem.

Axle holes in aluminium wheels tend to become enlarged in the course of time due to the hammering effect of vibration. Although such wheels

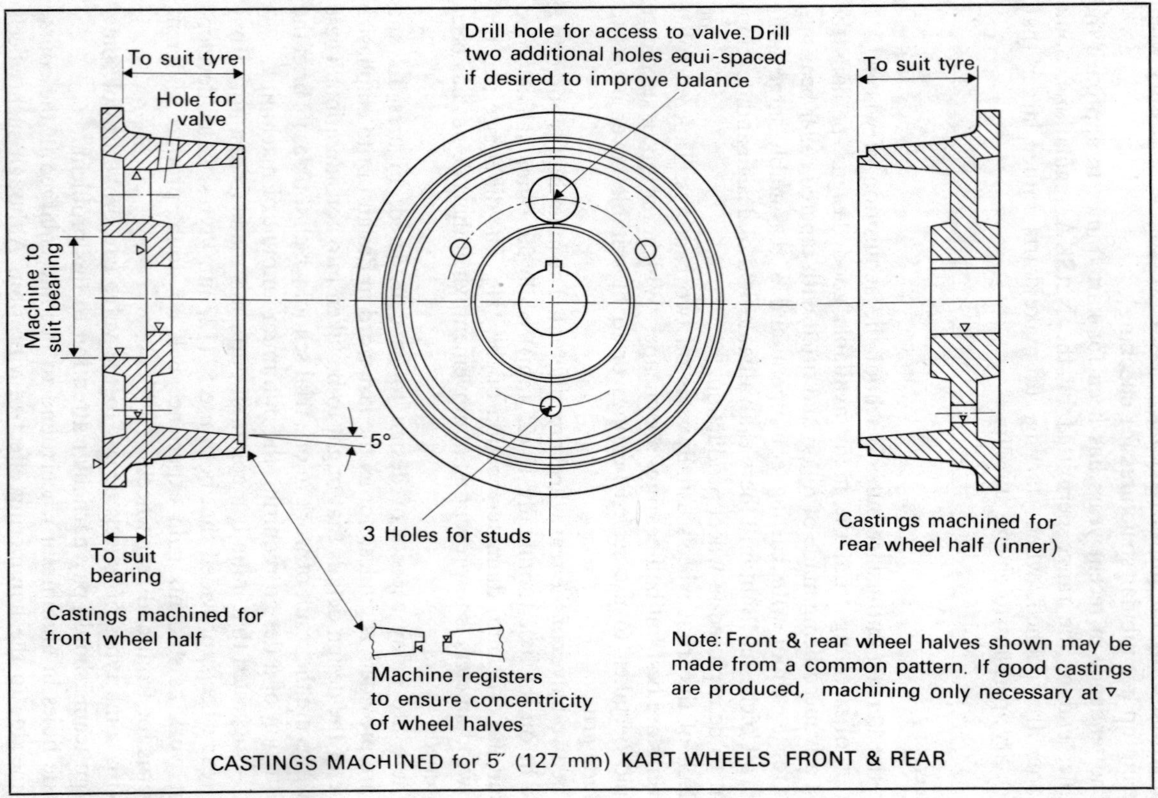

Drill hole for access to valve. Drill two additional holes equi-spaced if desired to improve balance

3 Holes for studs

5°

To suit tyre

Hole for valve

Machine to suit bearing

To suit bearing

Castings machined for front wheel half

To suit tyre

Castings machined for rear wheel half (inner)

Machine registers to ensure concentricity of wheel halves

Note: Front & rear wheel halves shown may be made from a common pattern. If good castings are produced, machining only necessary at ▽

CASTINGS MACHINED for 5" (127 mm) KART WHEELS FRONT & REAR

3.2 Wheel design drawing.

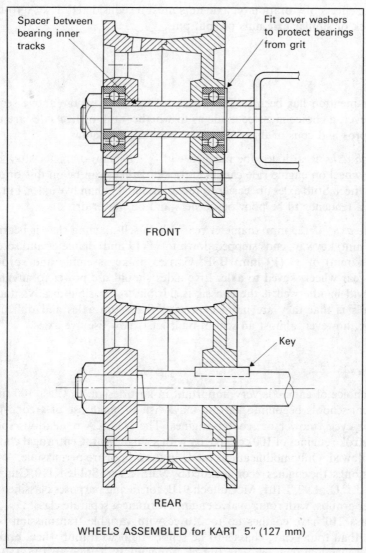

Spacer between bearing inner tracks

Fit cover washers to protect bearings from grit

FRONT

Key

REAR

WHEELS ASSEMBLED for KART 5″ (127 mm)

3.3 Wheel design drawing.

will last for a season or two, more sophisticated designs with steel central inserts might be considered.

If wheels are to be purchased Connolly or Trokart steel wheels are recommended. They are more robust than alloy or aluminium wheels. If only for school-yard use, scooter wheels will do. But avoid solid tyred wheels. It is best to standardize at 3·5″ (89 mm) × 5″ (127 mm) tyres (Avon kart slicks and Motorkart tubes).

25

The effects of vibration on axle nuts must be considered. Use Aero-tight, Nylock or similar axle nuts or split pins.

Axles

Some mention has been made of axle requirements in the above section. However, a choice must be made as to whether stub or live axles are used. The pros and cons of these two types are outlined below.

Stub axle. Rear axle tube plugged with $\frac{3}{4}''$ (19 mm) diameter stub axles. Rear wheel on engine side carries sprocket. Brakes can be on this or other rear wheel. Stub axles are easier to repair and make than live axles, but they have a tendency to be poor on corners and on power drive.

Live axle. $1''$ (25 mm) diameter rear axle in self-aligning chassis bearings; $\frac{1}{4}''$ (6 mm) keyway, ends stepped down to $\frac{3}{4}''$ (19 mm) diameter and screwed $\frac{5}{8}''$ (16 mm) or $\frac{3}{4}''$ (19 mm) BSF. Carries brake assembly and sprocket, with rear wheels keyed to axle. Live axles should use power to advantage and will handle well if the chassis is flexible. A possible disadvantage in schools is that they are more expensive than stub axles and difficult to repair; however, almost all school-built karts now use live axles.

Engines

The choice of engine is very important in kart design. In Class 100 (up to 100 cc) schools beginning karting may wish to make use of second-hand motor cycle, mower or scooter engines. The NatSKA regulations specify two-stroke engines of 100 cc capacity with a single gear. Centrifugal clutches are allowed. Only modifications made in the school are permissible.

Amongst the engines recommended by NatSKA are Stihl SK110, Guazzoni SV9, J.L.O., L99, L101, McCulloch 91B, for racing purposes classified into power groups, with rotary valve engines forming a separate class.

Class 210 have engines up to 210 cc with variable transmission of no more than four gears, chassis to be school built with four wheel braking. Engines lending themselves to this class include BSA Bantam 125, 150, 175 and Villiers 6E and 8E. This will necessitate design provision of clutch pedal and gear change lever.

Steering

Discussion of steering geometry in enthusiastic karting circles tends to become both complex and heated. It would appear that each variation in castor angle, kingpin inclination, and so on, has its own devoted band of followers. But for school karting a knowledge of the basic factors involved should be sufficient. These are best communicated in diagrammatic form; the reader's attention is directed to figures 3.4 to 3.9.

Ergonomic aspects

It is quite a good plan to sit one student near the centre of a full-size layout of the kart and consider the ergonomic aspects of the design problem. Where will the wheels be placed? the controls? the engine? Nearly all suitable engines are air-cooled and should not be shrouded from the flow of air, but neither should they be placed where the driver's arm could be burnt.

Allowance must also be made for short and tall drivers. Variations in the driver's leg lengths tend to be taken up by the taller individuals bending their knees more. It is quite usual for the steering wheel to be between the driver's knees.

3.4 Principle of Ackerman steering.

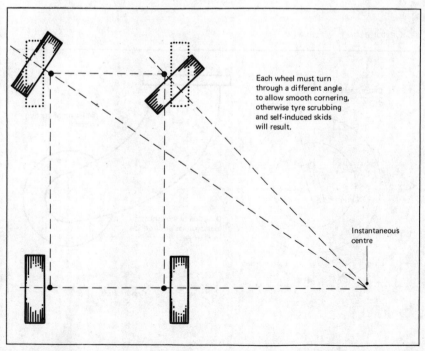

Each wheel must turn through a different angle to allow smooth cornering, otherwise tyre scrubbing and self-induced skids will result.

Instantaneous centre

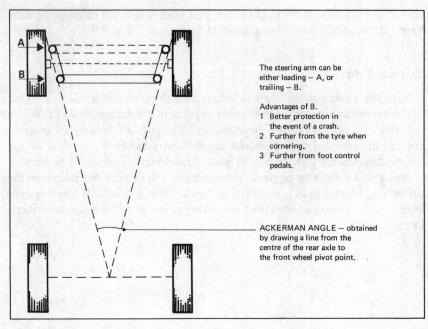

The steering arm can be
either leading — A, or
trailing — B.

Advantages of B.
1 Better protection in
  the event of a crash.
2 Further from the tyre when
  cornering.
3 Further from foot control
  pedals.

ACKERMAN ANGLE — obtained
by drawing a line from the
centre of the rear axle to
the front wheel pivot point.

3.5 Ackerman steering geometry.

3.6 Castor action.

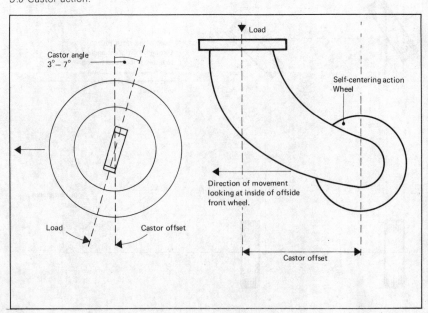

Castor angle
3° − 7°

Load

Self-centering action
Wheel

Direction of movement
looking at inside of offside
front wheel.

Load   Castor offset

Castor offset

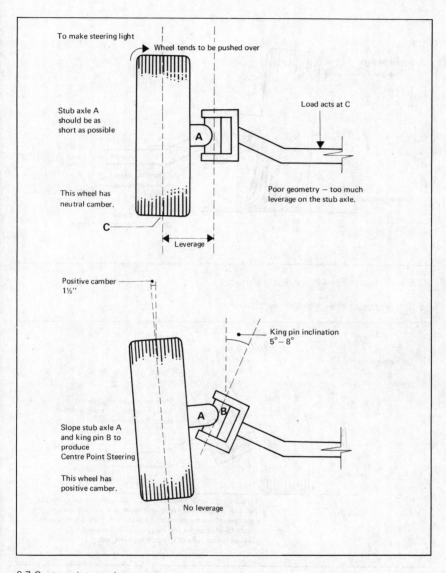

To make steering light

Wheel tends to be pushed over

Stub axle A should be as short as possible

A

Load acts at C

This wheel has neutral camber.

Poor geometry — too much leverage on the stub axle.

C

Leverage

Positive camber 1½"

King pin inclination 5° − 8°

A B

Slope stub axle A and king pin B to produce Centre Point Steering

This wheel has positive camber.

No leverage

3.7 Centrepoint steering.

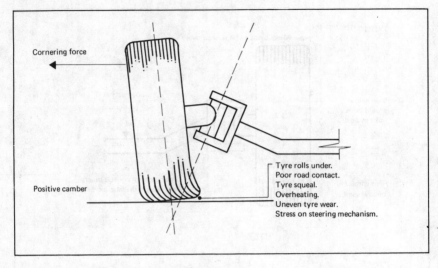

Cornering force

Positive camber

Tyre rolls under.
Poor road contact.
Tyre squeal.
Overheating.
Uneven tyre wear.
Stress on steering mechanism.

3.8 Steering geometry—theoretically correct.

3.9 Steering geometry—a better practical layout.

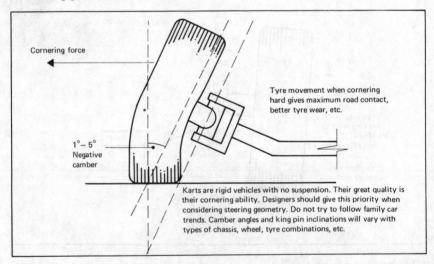

Cornering force

1°− 5°
Negative
camber

Tyre movement when cornering
hard gives maximum road contact,
better tyre wear, etc.

Karts are rigid vehicles with no suspension. Their great quality is
their cornering ability. Designers should give this priority when
considering steering geometry. Do not try to follow family car
trends. Camber angles and king pin inclinations will vary with
types of chassis, wheel, tyre combinations, etc.

Noise

Where karts are used on school playgrounds in residential areas the noise
problem could, at worst, result in the activity being stopped. Trouble
may, however, be avoided if a public relations campaign is carried out to
explain to the neighbours what is happening and why. It is, of course,
best to limit karting to reasonable hours and make it clear at the outset to
those living nearby that this is being done. A suitable silenced engine need
not make any more noise than a motor scooter.

The whole issue of noise measurement can be used as basis of study for the more technologically aware students. The design and manufacture of a simple noise level meter is within the scope of a science sixth former.

Aesthetics and kart design

Examination of the filmstrip complementary to this booklet will show that karting, for all its emphasis on function, offers a good deal of scope for the application of aesthetic design considerations. One kart, for instance, has a particularly attractive floor tray. Others have glass-fibre seats that are both comfortable and pleasing in appearance.

Some school karts, on the other hand, leave a lot to be desired in this respect. While it is dangerous to try to deal with an issue of this nature in a few words, it will be found useful to concentrate on simplicity and to ensure that the various shapes which come together to make up the finished job have been considered in relation to each other.

General considerations

It is sometimes suggested that, at the outset, a second-hand kart should be obtained at reasonable cost. This can be a useful motivating factor during the early design and construction stages as well as providing a source of information and experience. For beginners a low-powered but reliable engine is more important than a highly attractive chassis.

Once a group of students have designed their first kart they should have a very sound basis for further work. Their evaluation of one kart leads naturally to a design brief for the second.

Clearly, students will need access to plenty of technical information in the early stages of their development. A number of publications and other sources of information that schools have found to be useful are listed in the bibliography.

4
A Case Study

This case study describes the introduction and development of karting in a coeducational comprehensive school. Most of the work was carried out as an out-of-school activity, although some of the design and construction of karts took place during school hours.

A first kart

The school's involvement began with a small group of third-year boys who wanted to make a kart. They began by looking at professionally designed karts which the school was able to borrow from local enthusiasts. This gave the students their first experience of the importance of such design features as construction methods, layout of parts and steering geometry.

4.1 A first kart.

The first kart owed a great deal to the professional examples, although in translating the ideas to suit the materials and facilities available in the school workshop the boys had to face a number of new design problems associated with fabrication. Their aim was to reproduce the layout used by the professionals, but since the only available bending machine was a simple device with limited application they had to reconsider the bending, welding and ease of assembly of the chassis. The engine for the kart was obtained from a lawnmower.

Evaluation and re-design

In performance the kart was rather underpowered and a little sluggish in handling. It had a tendency to pull towards the right, steering nicely into right-hand bends but reluctant to turn into left-hand bends. Our knowledge of steering geometry was limited and at the time we could not find a book on the subject simple enough for our needs*.

Clearly, steering was an important area of design development to be tackled in the Mark II version. In addition to this the students wanted a more powerful engine. This desire for greater speed made them conscious of the fact that the kart's handling and general construction had to be improved, for safety's sake, and this factor influenced much of their thinking.

Development work was aided by the fact that the first kart crashed. The students rebuilt it, speculating on what might have caused its poor handling. They decided that the key to the problem lay in the way the weight was distributed.

On the Mark II kart the parts were rearranged to spread the weight more evenly. For example, an attempt was made to counterbalance, in part, the weight of the driver by the weight of the engine. In this way the students brought the centre of gravity to the middle of the kart, whereas commercially built karts have their centre of gravity close to the back wheels. The arrangement worked well and resulted in greatly improved handling.

The students also discovered at this stage that they needed a simpler steering geometry. It was apparent that their level of design thinking was deepening and that their decisions were becoming less dependent on the precedents they had examined initially. At the same time they began to develop simpler versions of certain components—brake pedals, for example—and to be more deliberate in their investigations of the position of components such as the steering wheel and foot-pedals where ergonomic considerations had to be taken into account.

*The basis requirements of a steering system are outlined on pages 26–30 of this book. For a more detailed discussion of the subject see Kart Design and Construction.

4.2 Mark II kart. *Note*: regulations have changed since this photograph was taken; these now provide for the wearing of securely fastened protective clothing and the wearing of goggles and gloves.

It should be noted that these investigations were based largely on observation and experiment rather than on theory.

Design organization

At the outset the intention was to break down the design of the kart into its main components—chassis, engine, wheels, steering, and so on—and to have a separate group of students develop each component. In practice this did not happen, partly due to fluctuations in karting club attendances and partly due to the fact that, in the absence of direction from the teacher, the students tended to gravitate towards groups of like rather than mixed abilities.

As a result of this process of self-selection some groups would comprise a number of able students capable of contributing ideas across the whole spectrum of kart design. Having solved the problems related to their own component they became redundant and inevitably began to influence other groups who were still grappling with their problems. They would observe a group's discussions, recognize the essential aspects of the problem and propose a solution. Often these suggestions were seen by the group to be 'right', or at least worth pursuing, and were adopted by them.

Thus the development of karting has been influenced by a small nucleus of able boys. But even within this nucleus there is a hard core of very able students with a sound basis of knowledge and experience of karting.

The range of design problems

The range of problems encountered in designing a kart is enormous, both in breadth and depth. Every single component could, for example, be considered as a challenging design problem in its own right. In practice, of course, a teacher must examine the range of components involved and decide that some are worth tackling as developmental design problems while others are not.

The chief criteria applied in this situation are those of complexity and cost. A sprocket carrier, for example, is fairly expensive and is quite a sophisticated clamping device. It might be considered unnecessarily sophisticated in its present form and its re-design would provide a considerable challenge.

Other components which appear to have considerable design potential include steering linkage and stub axles; both allow a wide variety of workable arrangements. All components, whether bought in or specially designed, should be examined to see how well they function and how well they fit into the overall design of the kart. Such studies may frequently lead to further design work.

Levels of design activity

Most problems encountered in kart design are of a developmental nature—concerned with modifying and refining what already exists. But there are areas, too, where the problems necessitate a real degree of new thinking. One such area—described below—lies in the adaptation of a brake unit. Other fruitful sources of problem-solving situations include the design of engine mountings sufficiently versatile to allow transfer of an engine from one kart to another, and the design of certain specialized components.

A brake unit conversion

The conversion was undertaken by a fifth-year group which fluctuated between one and four boys of average ability. The work took three weeks, including twelve timetabled periods and a few extra periods together with some evenings and parts of weekends which they worked because they wished to do so.

The original brake unit was obtained from a scrapped Lambretta scooter and the boys quickly saw that it could be altered to fit the kart. The problems they faced were new—they had never converted a brake unit before, although they were familiar with basic machining processes.

No formal analytical work was done but in joint discussion between teacher and students five aspects of the conversion were gradually established :

1 to remove the heavy and bulky swing arm to which the brake unit was attached ;
2 to reduce the weight of the unit by machining away all extraneous parts ;
3 to cast an auxiliary plate on to one side of the unit and make provision for the fitting of a special hub unit ;
4 to design and make spacer lugs to support the auxiliary plate against the bearing mounting ;
5 to carry out all the work without touching the internal brake face which was to be used as a datum for all measurements.

The discussion of these requirements and their empirical examination was strongly supported throughout by sketches drawn on paper, blackboard and floor, and the work carried out directly on the unit.

4.3 The converted brake unit.

The conception of the conversion and its planning and execution were entirely directed by the students. The work involved them in a fairly high level of developmental design. At the conclusion of the job the degree of accuracy obtained was within 0·0007" (0·0179 mm) and the completed brake was entirely successful in operation.

Karting for junior students

The original kart club started with third-year students, but as they progressed through the school to the sixth form no one came forward from the junior years to take their places. To promote a flow of new talent and to establish karting as a developing activity through the school a junior club was created for second- and third-year students. The club's first kart was the crashed and now rebuilt lawnmower-engined version handed down from the senior club.

First experiences

Knowledge of karts and karting is provided first of all by the actual experience of driving the club kart. Problems of weight distribution, handling, safe cornering, power, positioning of components, and so on, are realized—at first dimly—by the effects they produce on the comfort, safety and efficiency of the driver. And the enjoyment the students get from driving is the greatest single motivating factor for further work.

If they are to drive the kart the students must also be capable of keeping it in running order. Maintenance, while not being taught in any formal way, does instil familiarity with basic components and sub-components and their purposes, as well as the need for following certain procedures and routines. As such, it is a valuable basis for the design work that follows.

Early design activities

The focus for the students' karting experiences is a track meeting held most Saturday mornings on the school tennis court. The meeting lasts for four hours and at five-lap intervals certain operations must be performed on the kart. In order to ensure that this happens a group of second-form students—acting entirely on their own initiative—fitted a mechanical device to the engine which shorts out the ignition system after five laps.

Their attitude to this cut-out device, which is shown in many other situations on track day, is an interesting reflection of their

4.4 Maintenance work on a school kart.

priorities : they will resort to any degree of making-do in order to keep the kart running. Everything else is subordinated to their determination that the kart should not be off the track any longer than necessary. On one occasion, when the cut-out worked loose—the teacher had deliberately given no advice about fixing—the students were forced to reconsider the problem. The Mark IV version proved capable of functioning throughout the entire four-hour session without attention. With only the minimum teacher intervention the enthusiasm generated by driving and watching the kart is channelled to overcome the problems which arise.

Designing a first kart

The students' next step is to apply their knowledge and experience by designing and building their first kart. They start by examining a number of the school karts to compare and evaluate the merits or faults of their construction.

This work, coupled with their previous karting activity, will enable them to produce a list of all the components their kart must incorporate.

This list is then divided into two sections : those parts to be bought and those to be made. The second section is further subdivided into parts to be made using standard moulds (wheel hubs, for example) and parts to be designed afresh and made from new patterns (the steering wheel, for example).

4.5 A wheel hub made in the school workshop using a standard mould.

4.6 A specially designed steering wheel. *Note*: regulations have changed since this photograph was taken and the bumper shown here does not now comply with chassis specifications.

The policy here is that if the standard pattern proves, on examination, to be as good as or better than the juniors could hope to achieve at their present stage then it is used. If not, a new design and mould are made.

Organizing the design work

When the parts lists have been prepared the students are grouped into design teams, each responsible for investigating one part of the total problem. The school's experience suggests that this method of organization is the one likely to produce the best results. The teaching pattern is, again, a mixture of formal and empirical methods.

Design of the chassis begins in an empirical way. This basic part of the kart provides an opportunity for an examination, by means of scale models, of the pros and cons of various positions for the other major components. Ergonomics are obviously involved here; one particular aspect that may be explored is the possibility of providing some method of changing the position of the seat, the steering wheel or the foot-pedals so that the seat-steering wheel distance or the seat-foot-pedals distance can accommodate a reasonable variation in the size of drivers.

4.7 Although school karts may need to accommodate driver size variations, the fact that knees can be bent to suit the seat–pedal relationship does offer some flexibility. *Note*: regulations have changed since this photograph was taken and complete gloves (not mitts or open backs) must now be worn.

Aesthetic aspects

The aesthetic aspects of design are seen to have a place in karts, in the abstraction of shape and form (particularly where identifying colours and markings are involved), and also in the sense the craftsman is familiar with—neatness, perhaps even elegance, in the arrangement of parts, careful workmanship and a good surface finish.

It is interesting to note that if asked to design the layout of a kart junior students will take their decisions, initially, not on the grounds of what functions best but on aesthetic grounds—what looks best.

4.8 A well-designed kart built by two CSE students.

5
Filmstrip:
Kart-ways*

This filmstrip has been designed to serve two functions. First, it is intended that it should provide a basis for an introductory talk to a group of students who have little knowledge of karting but who plan to design and build their own kart. Here, the filmstrip's purpose is to motivate as well as inform. Secondly, many of the illustrations may be used as starting points for discussion during the actual process of designing.

The filmstrip may be considered to be divided into two distinct, but complementary, sections. Frames 1–8 and 33 deal with the activity of karting; frames 9–32 deal with kart design and construction.

Note: regulations have changed since this filmstrip was produced and attention is drawn in particular to the following:
Frames 3–7: these show a track marked out with tyres laid singly. Experience has shown this to be unsatisfactory and current recommendations are that, where tyres are used, they should be tied together in vertical columns of three's.
Frame 4: regulations now provide for the wearing of securely fastened clothing and the wearing of goggles and gloves.
Frame 20: the bumper shown here does not now comply with chassis specifications.

*Notes and Commentary to accompany Design and Craft Education Filmstrip No. 16.

Commentary

KARTING

Introduction—clothing

1

Karting has been called the fastest growing motor sport in the world. It has its own clubs, its own tracks and race meetings, even its own world cup! Anyone who has driven a kart will realize why the sport's followers are so enthusiastic.

A kart appears to be a fairly simple vehicle; little more than four wheels and an engine. Of course, it's not quite as simple as that, as we shall see later. But many schools have designed their own karts to their own specification.

This boy is driving a school-built kart. Note his protective equipment— helmet, gloves, eyeshields and suitable clothing.

Track layout

2

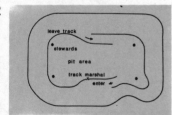

Here is the layout of the track he is using. Before a driver enters the track he will wait for the track marshal to signal him on to it, and will ensure that he does not pull out in front of another kart.

On the opposite side of the track we can see the lane a driver will use when he wishes to leave the track—raising his arm to signal his intention. If a kart should stop on the track for any reason the track stewards would run to the driver's assistance and quickly lift the kart back into the pit area.

43

3, 4 Track construction

3

This picture shows a kart entering the track. The course is marked out with old motor tyres. These stop a kart if the driver loses control . . .

4

. . . at a tight corner, for example. There seems little danger of that happening here. Although, in view of the position of the kart in the foreground, the driver at the rear should perhaps be ready to take avoiding action.

Track behaviour

5

If a kart strikes a tyre, or any other obstruction, and stops, the driver must switch off his engine and wait for the assistance of the track stewards.

Track stewards

6

Two track stewards can be seen at their posts in the background here, wearing white helmets. The karts in the foreground don't seem to need any help at present!

44

Track behaviour

The driver raises his arm to show that he is about to leave the track for the pit area. He must switch off his engine at the end of the run in, and not drive around in the pits.

Maintenance

Any kart, however well designed and constructed, will need some work done on it from time to time. In any case, regular checks must be made on machines during the course of a track meeting.

KART DESIGN AND CONSTRUCTION
Introduction—square-tube frames

We have seen something of karts in action; now let's look at how they are constructed.

A kart consists basically of a frame with a wheel at each corner and holding an engine, fuel tank, steering wheel, brake and accelerator pedals, and a seat for the driver.

That definition is, of course, not a design specification but a starting point. For example, one thing we need to know when designing a kart is how big the frame should be, what shape and of what material it should be made.

This is a rather unusual frame made from square section tube. You will notice that one piece has been welded above another to give extra strength.

Round-tube frames— no tube bender required

10

The use of round tube is more usual. This frame, however, has not required the use of a tube bender.

Round-tube frames— tube bender required

11

Here is a more conventional frame, needing a tube bending machine as well as welding facilities.

General example—as examination course work

12

This is a beautifully built kart. By the way, two boys successfully entered it as their examination course-work. Not surprisingly, they did well!

Wheels and tyres

13

Many schools have also designed and cast their own wheels. Arranging for access to the valve needs special thought.

This tyre is called a 'slick', because it has no tread pattern. The depressions in the face of the tyre are provided to indicate whether sufficient rubber is left for effective use.

Braking—drum brakes

Brakes are vital, of course. Here is a typical drum type. Class 100 karts—those with maximum engine capacity of 100 cc—only need one on the back axle. It is keyed on to the axle to prevent rotation.

Braking—twin drum brakes

In some circumstances more than one brake is fitted. This arrangement consists of twin drum brakes with a compensator.

One of the functions of the Kart Scrutineer who attends all track events is to check the brakes of the karts taking part. He does this by standing in front of the kart and pulling the brake pedal towards him. If the kart moves it is not allowed to enter the track.

Braking—disc brakes

Disc brakes are sometimes fitted. These are usually very efficient and make a very interesting design and engineering project. Such a brake may be mechanically or hydraulically operated.

Steering

Steering provides interesting problems. Here we see one arrangement which allows the front wheels to pivot under control of a track rod.

Steering and foot-pedal controls—floor tray

18

Some considerable thought and care has obviously gone into the construction of this steering and foot-pedal control system. Note, too, the floor tray. Workmanship of this standard never fails to impress. Clearly kart construction involves more than metal-work!

Steering and foot-pedal controls

19

Again, beautifully engineered controls— all designed and built in a school workshop.

You may wish to refer to these slides when working out your detailed design problems, not to copy a particular arrangement or mechanism, but to compare ideas.

Steering and foot-pedal controls— ergonomics

20

This picture shows the entire steering and foot control layout. When designing your steering system remember to consider the driver. The study of the physical relationship between machines and their human operators—called ergonomics—is an important part of design.

The controls of a kart must be placed where they can be operated comfortably by the driver. There is little point in placing a steering wheel 80 cm away from the seat if the driver's arms are only 60 cm long!

Seating

If the driver is to be safe and comfortable he will also need a well-designed seat. It may be made of upholstered metal or wood, or glass-reinforced plastics.

Engines—positioning

A wide range of engines is available to the kart enthusiast. This is a very reliable German JLO engine. The fact that a duct is fitted to convey cooling air to the cylinder finning means that the engine can be tucked behind the seat if necessary.

This is the kind of point which must not be overlooked in kart design. A hot engine too close to the seat can result in a most uncomfortable ride for the driver!

Engines—exhaust and chain guards

The Stihl engine shown here is lively and very suitable for school use. Note that the hot exhaust system is guarded to protect the driver's arm and leg.

Chains, too, must be guarded and particular care must be taken to ensure that the driver is protected should the chain break.

24, 25 Engines—adaptations

It is often possible to adapt an engine originally intended for a quite different purpose. This McCulloch engine was designed to drive a chain saw. It makes a very potent unit for a kart.

25

This kart has a less sophisticated engine taken from a lawn mower or cultivator. Such engines are excellent for first karts and are usually very reliable when in good condition.

Engines—axle bearings—silencers

26

Note the self-aligning bearing that supports the back axle. The silencer must not protrude beyond the frame.

Motor cycle and moped engines—RAC/ NatSKA Regulations and engine classification

27

You may recognize this as the familiar Villiers engine used on small motor cycles and mopeds. Very useful for anyone just starting karting, these engines have a reputation for reliability.

If you intend to use your kart at track events careful attention will have to be paid to the RAC or NatSKA (National Schools Karting Association) Kart Regulations. These set out certain minimum standards of kart design. They also classify karts according to engine capacity and the type of transmission.

So, for example, if you are adapting a motor cycle engine for kart use you will have to check first to see what class it could be used in.

Engines—drive sprockets

This picture indicates the size of the rear sprocket needed if direct drive from the engine crankshaft is used. Here the drive ratio is 9:74 using a JLO engine.

Engines—positioning—chain guards

In this case the JLO engine is fitted to the opposite side of the kart, but still with the carburettor facing backwards. You may have noticed that a leather strap has been used as a chain guard.

Fuel supply

The kart shown here is fitted with a very small engine suitable for beginners —about 50 cc. The fuel feed system is rather interesting. Fuel is pumped from a tank on the floor tray. Most school karts use a gravity feed system.

Kart trailers

A useful trailer for towing two karts. The coach in the background gives an idea of the enthusiasm for karting in some schools. It was hired to make the long journey to the track which is situated on a disused airfield.

General example

32

This workmanlike kart was developed by a school from two earlier designs. A Stihl engine is used again.

KARTING

A kart at speed

33

The blurring here is a result of speed, not camera-shake! This is a student driving a school-built kart during an inter-school track event, in spite of the weather. Would you like to join him?

6
Karting Questionnaires

Introduction for students

Karting has been called the fastest growing motor sport in the world. The first kart was built in California in 1956 and the sport was introduced to Britain in 1959. Now, karting takes place in almost every country in the world. It has its own clubs, its own tracks and race meetings, even its own world cup!

Anyone who has driven a kart will realize why the sport's followers are so enthusiastic. There is nothing quite like the sensation of speeding round a track a few inches from the ground in a small, open vehicle like a kart. But there is more to karting than just the sensation of speed and excitement. Many young people have experienced the added interest and enjoyment to be gained from designing, building and maintaining their own karts. After all, the best drivers are those who understand their machines. And what better way to come to understand a vehicle than by designing and building it?

Let's assume you know hardly anything about karting but are interested in building your own kart. Where do you start? Obviously, the first thing you must do is find out more about karts. There are several ways of carrying out your investigation. Specialized books and magazines are available. These can be extremely valuable, particularly if there is a technical point you wish to check or a source of supply you need to discover. But perhaps the best starting point would be to visit another school which has been engaged in karting for some time. Then you can take a careful look at the layout of their kart and, most important of all, find out WHY particular design decisions were taken. In discussion you can also find out the kinds of problems they encountered.

If you do not know of such a school it might be worth contacting the National Schools Karting Association (NatSKA). This association has the names and addresses of many schools engaged in karting and may be able to advise you if there are any within reasonable travelling distance of your own school. Some counties have their own karting associations who are pleased to offer advice and encouragement to newcomers to the sport.

If there are no karting schools near you, do not lose heart. There are plenty of other sources of help and information. And there is something to be said, too, for being a pioneer!

Planning for design

Anyone setting out to design, construct and drive a kart must plan their work very carefully and ensure that all the relevant design factors are taken into consideration. You will probably find that the best way to tackle this is to work in design teams, each team being responsible for one particular area of investigation. The work may be divided into eight such areas. These are:

1 the basic frame;
2 wheels and braking;
3 the engine—types and methods of mounting—the exhaust system;
4 axles—bearings—the drive mechanism;
5 steering;
6 the fuel system;
7 seating and controls;
8 social organization.

These eight areas of work are dealt with in more detail in Questionnaires 1 to 8. To help you in your investigations the various factors you will have to consider have been set out in the form of questions. Your answers to these questions will enable you to draw up a specification. Once this has been done it only remains for you to translate the specification into a kart! This may involve practising many of the practical skills that you have already learnt. Or it may mean that you need to learn new techniques—welding, for example.

In addition to the questions in the following pages, there are a number of general points that will need to be borne in mind by all the design teams. Perhaps the most important thing to remember is that the performance and safety of the finished kart will depend, first, on how well you carry out your investigation. You must therefore make sure that the information you collect is complete and accurate. Use any other publication or source of information to help you make your decisions.

At some stages it may be necessary for you to experiment with new materials or unfamiliar processes to discover for yourself the most acceptable solutions to problems. Do not hesitate to experiment in this

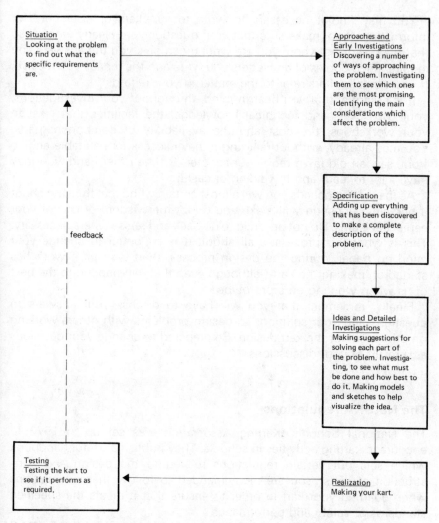

Situation
Looking at the problem to find out what the specific requirements are.

Approaches and Early Investigations
Discovering a number of ways of approaching the problem. Investigating them to see which ones are the most promising. Identifying the main considerations which affect the problem.

Specification
Adding up everything that has been discovered to make a complete description of the problem.

Ideas and Detailed Investigations
Making suggestions for solving each part of the problem. Investigating, to see what must be done and how best to do it. Making models and sketches to help visualize the idea.

Testing
Testing the kart to see if it performs as required.

Realization
Making your kart.

feedback

6.1 The design process. This is the kind of process you will follow when designing your kart.

way, and keep an accurate record of what you do and how you do it. This information may prove valuable at some future stage of the work.

To avoid confusion one member of each design team should be made responsible for keeping a record of the design work. These records should include your answers to the questions which appear in the following pages, together with any drawings or models that you make, or catalogues and letters that you collect. Where the work demands the co-ordinated efforts of several group members each one should make a note of the contribution that he made.

You may find it necessary to write to suppliers or elsewhere for information. If so, make sure that your questions are clearly worded so that the person to whom you are writing can give you the answers you require. Keep copies of any letters you write and the replies you receive. If you visit suppliers refer to the matter in your records.

It is essential that you bear in mind, throughout your investigations, your own skills and technical knowledge, the facilities that exist in your workshops, the cost and the availability of component parts. You may already, without realizing it, have access to a suitable engine from, say, an old lawn mower or scooter. On the other hand, you may have a lot to learn about welding or casting.

As the work proceeds you will almost certainly find that thinking about a particular question will make you doubt the wisdom of one of your earlier decisions. Do not be afraid to go back and revise where necessary. This is what this process is all about. It is far better to change your mind on paper, during the design process, than wait until work has started on making the kart—although even that can happen to the best of us when working on a prototype!

Finally, remember that you will have to discuss your answers to questions and your solutions to design problems with others working on other parts of the kart design. Be prepared to change your decisions as a result of your discussions.

The NatSKA regulations

The National Schools' Karting Association was set up in 1972 to encourage karting activities in schools. They publish a useful handbook which sets out certain regulations related to the design and construction of karts. You are recommended to follow these regulations when building your kart in order to ensure that it meets the specified standards of safety and performance.

The control of motoring events in the UK is organized by the RAC and a copy of their regulations will prove useful, particularly if you feel that you may wish to graduate towards out-of-school club activities. There is considerable overlap between their requirements and those of NatSKA; it would be quite feasible to design a kart which conformed to the regulations of both organizations.

Some schools have built first karts which did not conform to NatSKA or RAC specifications, particularly with regard to wheels or engines (scooter wheels, for example, are sometimes readily available). But it may be sensible to deviate from these regulations only in minor ways so that the kart may be made to conform to them at a later stage by means of simple modifications.

(1) The basic frame

1 What types of material are most suitable for use in constructing a kart frame in your own workshop?
   a Square steel tube? If so, what would be the best size?
   b Round steel tube? If so, what would be the best diameter and thickness?
   c Steel girder section? If so, what would be the section and thickness?
   d Are any other materials suitable? If so, name them and give brief details.
2 Which of the above materials have you decided to use?
3 From whom will you be able to obtain the material?
4 About how much will you require (of each size, if relevant)?
5 How much will it cost? (Write for a quotation if necessary.)
6 What types of joint construction are suitable for a kart frame?
   a Welding? If so, are facilities available at school? If not, is there a garage or factory locally where the work could be done?
   b Brazing?
   c Riveting?
   d Bolting?
7 What shaping processes are likely to be required?
8 What do the NatSKA or RAC regulations say about the design of the basic frame?
9 What are the size limits of the basic frame? (State the maximum and minimum lengths and widths.)
10 What types of joint will you use? If necessary make up two or three specimen joints to help you decide.
11 How do you intend to shape your material? Experiment if necessary.
12 Draw a plan of the frame and mark the dimensions.
13 How will you decide where to put the steering wheel?
14 How will you decide where to position the brakes and accelerator pedals?
15 Make a scale model of your frame in metal wire.
16 What extra struts are necessary to mount the engine?
17 What extra struts are necessary to support the seat?
18 What extra struts are necessary to support the steering column?
NB Keep all additional struts to a minimum in order to reduce weight, and the amount of work you have to do.
19 Does any special provision have to be made to support the driver's heels? If so, what? How will you provide this support?

NOTE: You may find it helpful to refer to the filmstrip Kart-ways during your investigation, particularly frames 9, 10 and 11.

(2) Wheels and braking

1 Are the wheels to be purchased? Or designed and made in school? (Questions *1–12* should be answered whether or not you intend to buy ready-made wheels. Questions *13–29* are only applicable if you are designing and making your own wheels)

2 What size of tyres will be used? NB You will need to specify the inside diameters of the front and rear tyres as well as their widths.

3 Where can these tyres be obtained?

4 How much will each cost?

5 How much will the inner tubes cost?

6 What size does the hole through the centre of the wheel have to be? (Give front and rear dimensions.)

7 Are bearings needed at the front and/or the rear?

8 Where can ready-made wheels be obtained?

9 How much would they cost? (Give front and rear prices separately.)

10 Does this include the cost of bearings?

11 Do you intend that your kart should conform to a particular NatSKA or RAC specification? If so, state which.

12 How does your answer to question *11* affect the design of wheels and tyres? Do your answers so far meet the specification?

13 If the wheels are to be designed and made in school: what materials are available? could the wheels be made from steel? wood? aluminium? plastics?

14 What processes and techniques would you have to use to make the wheel from the material of your choice?

15 Would it be possible to use a material not at present available in your workshop? If so, which?

16 Where could it be obtained?

17 How much would it cost?

18 Do you need any special equipment to work it? If so, is this equipment available in your workshop?

19 If it is not available, could it be made available?

20 To help you: if thinking about casting consider aluminium or zinc-based alloys. What are the advantages of each?

21 Are moulding boxes of sufficient size available?

22 Can you produce a pattern?

23 Have you a suitable lathe to turn the casting?

24 How can the casting be held on the lathe?

25 Which diameters on the wheel must be machined accurately to size?

26 Which diameters must be concentric?

27 How will you get access to the tyre valve when the wheel is assembled?
28 Most kart wheels are made in two halves. How will you assemble the halves?
29 How can the two halves be kept together?
30 How can the whole wheel be kept on its axle?
31 Do the front and/or rear wheels need to be prevented from rotating on their axles?
32 If this is necessary, how will you do it?
33 Karts tend to suffer from loose nuts and bolts. Why?
34 How can you prevent any screws, nuts or bolts from coming undone?

BRAKING
35 Have you adopted one of the NatSKA or RAC specifications? If so, which one?
36 What effect, if any, will your answer to question 35 have on your design for a braking system?
37 What types of brakes could you use?
38 Which types do you intend to use? Why?
39 Where can any necessary supplies be obtained?
40 How much will these supplies cost?
41 Will you need any special equipment in order to make the braking system?
42 How will you stop the fixed part of the brake from going round when the brake is applied?
43 How will the driver control the brakes?
44 How will you ensure that the moving part of the brake really does go round with the wheels?
45 What tests will the scrutineer at a track apply to your brake(s)?

REMEMBER to discuss your answers to all these questions with others working on other parts of the kart design. You should pay particular attention to your colleague's answers to questions 12–15 in Questionnaire 1 ('The Basic Frame').

NOTE: During your investigation you may find it useful to examine frames 13, 14, 15 and 16 of the filmstrip Kart-ways.

(3) The engine—types and methods of mounting—the exhaust system
1 What is the difference between a 2-stroke and a 4-stroke engine?
2 What is meant by the capacity of an engine?

3 Do you intend that your kart should conform to one of the NatSKA or RAC specifications? If so, state which.

4 How, if at all, will your answer to question *3* effect your considerations of 2- or 4-stroke engines? Or engine capacity?

5 Are you free to choose whether or not to use gears? A clutch?

6 What is meant by a centrifugal clutch?

7 Does your specification permit it?

8 What would be its advantages?

9 Discussion of questions *1–9* should have helped you to decide what kind of engine you are going to try to obtain. Will it be new or second-hand?

(Questions *10–13* apply to second-hand engines only.)

10 If second-hand what arrangements can be made for overhaul?

11 Which parts will you clean? How?

12 Which parts will you renew? How?

13 What tests will you perform when the overhaul is finished?

14 Where can you obtain an engine?

15 How much will it cost?

16 What kind of mounting arrangements will the engine need to have?

17 Could a motor cycle or scooter engine be modified to enable it to be mounted on a kart frame? How?

18 Which way will the carburettor face?

19 What kind of ignition system will it have?

20 Does it need a battery?

21 Will the ignition system be likely to fail if it gets wet?

22 How can a rear axle drive system be arranged?

23 Where will the engine be fixed with respect to the seat?

24 Would the driver be in danger of burning himself as a result of the way the engine is fitted?

25 What arrangements will be necessary for the exhaust?

Check your decisions so far against the current NatSKA and RAC regulations. Whatever specification you adopt these regulations provide expert advice to help you to use your kart in safety.

26 Does your design as formulated so far comply with these regulations?

27 Now, what have you decided to do about your engine? Make a summary of your decisions.

REMEMBER to discuss your answers to these questions with others working on other parts of the kart design. You should pay particular attention to your colleagues' answers to questions *12–16* in the 'Basic Frame' questionnaire.

NOTE: Frames 22–29 of the *Kart-ways* filmstrip may prove helpful during your investigation.

(4) Axles—bearings—the drive mechanism

  1 Do you intend that your kart should conform to one of the NatSKA or RAC specifications? If so, state which.

Even if you do not adopt one of these specifications you are recommended to study the current editions. In this way you may benefit from the suggestions of experienced karters.

  2 How will you support and drive the wheels?

  3 Which wheels will act as the drive wheels? The front wheels or the rear wheels?

  4 Do the rear wheels rotate separately? If not, what happens when cornering?

  5 What other parts must be associated with the axles? The drive mechanism? The braking mechanism? The axle supports?

  6 What arrangements can you make for the front wheels?

  7 How can you make the engine drive at least one wheel?

  8 How will you ensure that things intended to rotate with the axle do so?

  9 How can you stop the wheels falling off?

10 What can you do about friction (i.e., the rubbing together of moving parts)?

11 Where, in the design of a kart, is friction a help? And where is it a hindrance?

12 How can we arrange for the axles and wheels to revolve easily without undue wear?

13 How fast must the driven wheels go around in comparison with the engine?

14 How could you drive the wheels from the engine? By belt, shaft, gears or chain and sprockets?

15 Which of the techniques listed above will you use? Why?
(Your answers will depend on the results of the investigation work on engine types and mountings.)

16 Are you able to obtain the necessary parts and materials? If so, where from?

17 What is the cost of obtaining the above?

18 What processes will you need to use?

19 Are these possible in your workshop?

20 What kind of rear axle will be used? Live? Fixed with rotating wheels?

21 What will be the diameter of the axle?

22 How will it be supported?

23 State what arrangements you will make to drive one wheel or two wheels (state which).
(Before answering this question attention should be paid to the

results of the investigation of the 'basic frame'.)
24 Sketch out your final design for keeping the rear wheels on the axle.
25 Sketch out your final design for keeping the front wheels on the axle.
26 What kind of bearings will be used for the front wheels? And for the rear wheels?

REMEMBER to discuss your answers to all these questions with others working on other parts of the kart design. This is particularly important in the case of questions 15 and 23.

NOTE: You may find it useful during your investigation to refer to the Kart-ways filmstrip, particularly frames 28 and 29.

(5) Steering

1 How can we steer our vehicle?
2 What do the NatSKA or RAC regulations say about steering?
3 Do you intend to adopt these regulations?
4 Will you steer the front wheels or the rear wheels?
5 What kind of turning circle will you require? (In other words, how abruptly will you want to turn?)
6 Will the steered wheels strike the frame when cornering?
7 When cornering, will both steered wheels point in the same direction?
8 What do we mean when we say that the wheels are self-centering?
9 How can you arrange for the steered wheels to self-centre?
10 How will you ensure that the steering is not too 'heavy'?
11 How quickly do you want to be able to direct the steered wheels on a kart?

NB: These are very difficult problems. It is most unlikely that you will be able to solve them yourself without expert advice or studying available information. It is vital to the success of your design, however, that you understand the questions and why they must be answered.

12 In collaboration with your colleagues working on the basic frame, wheels and braking, and the axles, bearings and drive mechanisms, sketch out the general layout of the frame and wheels.
13 Sketch the general layout of the steering mechanism on the above plan.
14 Consider the driver in relation to the steering wheel. Where will the wheel be? How big will it be?

15 How will the steered wheels be supported? (Your colleagues responsible for designing the axles, bearings and drive mechanism should be able to assist you here.)
16 Will the steered wheels move together on one axle? Or on separate short axles?
17 If on short (stub) axles, how will each be fitted to the frame?
18 Will there be clearance on 'full lock' (i.e., when the wheel is fully turned)?
19 How will the wheels be connected to the steering mechanism?
You are advised to look again at some of the frames in the *Kart-ways* filmstrip considering these points, particularly numbers 18, 19 and 20.
20 Use full-size mock-ups to try out your ideas.
21 Finally, make a full-size detailed drawing of your design.
22 When the kart is tried out go back and see if you have successfully answered all the questions. Do not be surprised if it proves necessary to modify the design after trials.
23 Is all your steering mechanism protected from accidental damage by contact with the ground or with other karts?

REMEMBER to discuss your answers to all these questions with others working on other parts of the kart design. This is particularly important in the case of questions *12–15*.

(6) The fuel system

1 What do the NatSKA and RAC regulations say about fuel systems?
2 What safety hazards can you see, and what must you do to minimize them?
The following investigation will necessitate close co-operation with your colleagues working on the engine and exhaust system.
3 Does the engine you intend to use pump its fuel from the tank or does it require gravity feed?
4 What difference will your answer to the above question make to your design?
5 What special points must you bear in mind about:
   a containers for fuel?
   b lids?
   c paintwork?
   d filling?
   e pipes?
   f taps?
   g supports for tanks and pipes?
6 What would be the effect(s) of fuel spilt on the track?

7 Are people likely to push the kart along by pushing on the fuel tank?
8 How does your answer to question 7 affect your design?
9 How can the tank be emptied if necessary?
10 Is the feed pipe likely to get damaged?
11 Could the feed pipe become hot?
12 How would vibration affect your fuel system?
13 Can air get into the tank to replace fuel as it is used?
14 Considering the results of your investigation make a general sketch of the fuel system layout.
15 Check that your layout will satisfy all the points raised above and is in accordance with the work of your colleagues.
16 Make detailed drawings of all the parts to be made.
17 What materials will be needed?
18 Where will they be obtained?
19 How much will they cost?
20 What processes will you need to use?
21 Do you have facilities for these in your workshops?
22 Check again that all the factors revealed in your answers to questions 1–13 have been satisfied.

(7) Seating and controls

1 What do the NatSKA and RAC regulations say about the seating and controls?

Your investigation of this area of kart design will involve you in the consideration of ERGONOMICS, i.e., the study of machine design in relation to its human operator. You will realize that this kind of work is applicable to very many fields . . . from arranging furniture in a room (a machine for living, eating or sleeping in) to designing the controls of an aircraft. What you learn from this investigation will carry over into much other work that you will do, both at school and outside.

You will have to consult everyone else working on the design, because their work will affect yours. It is best to carry out this stage when either the basic frame is available or at least a full-size drawing has been made. Perhaps the best way to start is to get someone to sit in the appropriate position on the frame or the drawing. A look at the filmstrip *Kart-ways*, particularly frames 18–21, may also prove useful.

2 What will you make the seat from? Wood? Metal? Plastics?
3 How much will the material cost?
4 What processes will you need to use?
5 Do you have the facilities for these in the workshops?
6 Do you think that the finished appearance will be pleasing? Does this matter?

7 Can you consider the appearance of the seat alone, or the seat together with the rest of the kart?

8 How important is the driver's comfort? What arrangements are you making to take this into account?

9 Where will the seat be placed?

10 At what angle will it be?

11 Where will the steering wheel be placed?

12 Where will the foot controls be placed?

13 Can you protect the feet and controls from damage in the event of a collision? Do the NatSKA or RAC regulations have anything to say about this?

14 How will you go about providing a floor tray to prevent the driver's feet from touching the ground?

15 From what material will you make the floor tray?

16 Should this be considered in relation to the seat, with regard to materials? Appearance?

17 How much will it cost to make the floor tray?

18 What processes will you need to use?

19 Do you have the facilities for these in your workshops?

20 It is essential, for safety reasons, to have a cut-out switch for stopping the engine. Where will this be connected?

21 Where must the cut-out control be placed?

22 Is it possible to ensure that it won't be damaged?

23 Make a full-size layout showing the relative positions of:

   a the steering wheel;
   b foot controls;
   c seat;
   d floor tray;
   e guard rails and chain guard;
   f cut-out switch.

24 Is there any danger of the driver being burnt by the exhaust system?

25 Can the driver easily operate all his controls?

26 What happens if a small boy tries to drive the kart? Or a tall man?

27 Are all the controls protected from damage?

28 Make detailed drawings of all the parts to be made.

(8) Social organization

1 Who will be allowed to drive your kart?

2 Who will pay for fuel? New tyres? Repairs?

3 Who is responsible if there is an accident?

4 Whose fault is it if people living nearby complain about the noise?

These questions, and many similar ones, are concerned with the 'design' of the organization of the people involved. If you look up the definition of 'design' in the dictionary you will see that it is not just related to machinery and things. Just as the design of a kart involves investigating and solving interrelated problems, so does the design of a karting event. Although the problems are different they can be tackled in much the same way.

5 Human beings organize all kinds of groups to fulfil all sorts of needs. Here are some examples. Can you add ten more?
Tennis Clubs
Schools
Allotment associations
Prisons
Youth Clubs

The examples given—yours and ours—will vary widely. The kinds of organizations set up will depend on the job they have to do (just as the design of a kart is determined by its job). Here are some questions that will help you decide what kind of organization you will have to set up.

6 Karting is dangerous if done carelessly. Must an adult take the lead?
7 What is your headmaster's responsibility for such an organization? He might be willing to make an appointment with two or three of you to discuss the matter.
8 What is your teacher's position in law if there is an accident?
9 Will there be money involved in running karting in school? If so, where will it come from?
10 How will it be accounted for?
11 Should students be free to decide everything?
12 Who will lead the organization—an individual or a small group?
13 Who will decide how to run the meetings?
14 Who will write any necessary letters?
15 Will you need to write down rules?
16 Where will you run your karts?
17 Will you allow privately owned karts to be brought?
18 Can you get permission to use the site?
19 Who will lay out the track?
20 Who will clear up afterwards?
21 How many people will there be in the organization?
22 Can *anyone* drive the kart(s)?
23 Can parents, friends outside school or people who have left school come and drive?
24 Is insurance necessary?

When you think you have satisfactorily answered all the above questions, make notes about the aims, organizational details and officers (if any) that you will need. This will be a *draft constitution*.

25 How can your draft constitution be given any authority?
26 How can it be changed later if necessary?

Bibliography and Further Reading

Burgess, A T. *Starting Karting*. Karting Bookshop.
Burgess, A T. *Kart Design and Construction*. Lodgemark Press, 1970.
Kart Plans Sets Numbers One and Three. Lodgemark Press.
Karting. A monthly magazine for kart enthusiasts.
McGregor, J. *Karting for Schools and Youth Groups*. Lodgemark Press.
Royal Automobile Club. *RAC Kart Regulations and Fixture List*. Published
annually by the Royal Automobile Club, 31 Belgrave Square, London
SW1.

All the above publications are available from Karting Bookshop, Bank
House, Susan Wood, Chislehurst, Kent.

Starting Karting and *Karting* magazine contain the names and addresses
of most of the major UK kart dealers. Dealers' spares lists and catalogues
can provide a useful source of information for schools, one of the more
comprehensive is the *Blowkart Do-It-Yourself Catalogue* available from
J. J. Blow Ltd, Oldfield Works, Chesterfield, Derbyshire.

The National Schools' Karting Association (NatSKA) has been set
up to co-ordinate karting activities in schools, from the design, development and construction of karts to the organization of track events. A
handbook of karting regulations and specifications is available and other
publications are planned. For full details write to the Hon. Secretary,
Mr K. Breach, The John Warner School, Stanstead Road, Hoddesdon, Herts.

The teacher will probably also find it helpful to refer to other publications in the Design and Craft Education series. These include:

Schools Council. *Design for Today*. London, Edward Arnold, 1974.
Schools Council. *Materials and Design: A Fresh Approach*. London,
Arnold, 1974.
Schools Council. *Education Through Design and Craft*. London, Edward
Arnold, 1975.

Schools Council. *Looking at Design*. London, Edward Arnold, 1975.
Schools Council. *You Are a Designer*. London, Edward Arnold, 1974.
Schools Council. *Connections and Constructions*. London, Edward Arnold, 1975.